I is a Strange Loop

Marcus du Sautoy is the Simonyi Professor for the Public Understanding of Science and Professor of Mathematics at the University of Oxford and a Fellow of New College. He is the author of seven books, including *The Music of the Primes* and his latest book *Thinking Better: The Art of the Shortcut*. He has presented numerous radio and TV series, including a four-part landmark TV series for the BBC called *The Story of Maths*. He works extensively with a range of arts organisations bringing science alive for the public, from the Royal Opera House to Glastonbury Festival. He received an OBE for services to science in the 2010 New Year's Honours List, and was made a Fellow of the Royal Society in 2016.

Victoria Gould studied at Manchester and Cranfield Universities, gaining a BSc (Hons) in Physics and an MSc in Applicable Mathematics, and also trained at L'École Jacques Lecoq Laboratoire Etudes du Mouvement. She is an Associate with Complicité, having collaborated with Simon McBurney on *The Elephant Vanishes*, *A Disappearing Number*, *Shunkin*, *Endgame*, *The Magic Flute* and *The Encounter*, with Annabel Arden and Clive Mendus on *Lionboy*, and with Poppy Keeling conceiving and developing their unique Ensemble Maths and Embodying Maths education projects. Her TV credits as an actor include two years playing the fearless journalist Polly Becker in *EastEnders*.

MARCUS DU SAUTOY
and
VICTORIA GOULD

I is a Strange Loop

faber

First published in 2021
by Faber and Faber Limited
74–77 Great Russell Street
London WC1B 3DA

Typeset by Brighton Gray
Printed and bound in the UK by CPI Group (Ltd), Croydon CR0 4YY

Marcus du Sautoy and Victoria Gould are hereby identified as authors
of this work in accordance with Section 77 of the
Copyright, Designs and Patents Act 1988

A CIP record for this book
is available from the British Library

978-0-571-36073-4

2 4 6 8 10 9 7 5 3 1

Preface

The play is divided into two acts which we have called 'The Mathematical Dimension' and 'The Theatrical Dimension'. The play grew out of our recognition that the two realms share much in common: establishing a set of axioms or rules that frame a hypothetical setting, followed by the exploration of the implications of these axioms. The rupture that happens as we go from the mathematical to the theatrical dimension reveals that the first act is an artificial creation that begins to bleed into the system that contains it. Yet which is the more real turns out not to be such an easy question to answer. There is a constant tension between what is possible in the mind and what is physically possible in reality, a tension that pervades both mathematics and theatre in its relation to the physical world around us.

Acknowledgements

I is a Strange Loop has been many years in the making and has evolved and grown, mutated and changed, in many different directions ever since we stepped into the rehearsal space together in 2010 to fan the first spark of an idea.

Devised theatre is by its nature a hugely collaborative process, to such an extent that it becomes difficult in the end to be able to truly identify the author. Brian Eno talks about the idea of 'scenius' rather than genius to indicate that any work of art has a whole host of people who contribute to its creation. It is therefore a pleasure to acknowledge those who were part of our journey.

The most important person who contributed so much to this piece is our director, Dermot Keaney, who has been part of the creative process almost from the beginning. Devising theatre means that dialogue and ideas emerge from the playful process in the rehearsal space, and a director is as important as the actors in creating the framework for ideas to emerge. His guidance, suggestions and contributions run all the way through the final script that our process converged on.

Other theatre practitioners who shared the rehearsal room with us at various moments during our journey have also played an important role. They include Simon McBurney, Joyce Henderson, Clive Mendus and James Yeatman.

We are also immensely grateful to those who contributed their production talents to bring the play alive. In particular: Stephen Hiscock (original music composition), Sean Ephgrave (sound design) and Christopher Nairne (lighting).

Several organisations had faith in the early stages of the piece and gave us support. Complicité, in particular producers Judith Dimant and Poppy Keeling, helped us to stage the very first performances at Secret Garden Party and Latitude Festival. We were very lucky that Jonathan Newby was there, in the tiny tent we performed in at Latitude, and then invited us to continue our development at the Science Museum. This culminated in the museum providing very generous financial support to the extent of even creating a bespoke theatre for us to perform in one of the museum's galleries.

Since then we've performed the play as it continued to evolve at numerous venues including the Glastonbury Festival, the Vaults Festival, the Hay Festival, the Manchester Science Festival and the Oxford Playhouse, as well as internationally in India and Spain. A generous grant from the Charles and Lisa Simonyi Fund for Arts and Sciences helped us bring the play to the Barbican.

It was the performance in The Pit at the Barbican in March 2019 where we felt that the play had finally evolved to a point that it was ready for publication. The play was staged there as part of the Life Rewired season. It was also here that the play finally received its definitive title: *I is a Strange Loop*.

We'd like to thank all those venues who over the years had the faith to stage a play about maths!

We would like to thank Steve King and all at Faber for all their work in transferring the play from stage to page.

I is a Strange Loop premiered in The Pit at the Barbican, London, on the 21 March 2019, as part of the Life Rewired season. The cast was as follows:

X Marcus du Sautoy
Y Victoria Gould

Director Dermot Keaney
Composer Stephen Hiscock
Sound Designer Sean Ephgrave
Lighting Designer Christopher Nairne

Variables

X
Y

I IS A STRANGE LOOP

'Through the unimaginable fourth dimension a man held captive in a prison cell could escape without passing through the ceiling, the floors or the walls.'

Jorge Luis Borges

Performance Notes

The script is based on the performance in The Pit at the Barbican. At a number of points it refers specifically to that location. For example throughout the play there is a recurrent sound of a train like the rumbling of the London Underground beneath the theatre. A different sound effect can be chosen in other venues, such as an offstage siren of an ambulance. When Y has to explain the sound to X an alternative definition can be substituted. Similarly when Y is describing the outside of the theatre an appropriate local description should be substituted.

The script also reveals the name of the actor playing Y. The actor playing Y may substitute any name of their choice for 'Victoria' in the script.

A more detailed description of the two Platonic Sequences performed by X at the beginning of the play can be found in the Mathematical Prompt Book that follows the play.

Act One

THE MATHEMATICAL DIMENSION

On stage we see the frame of a large white cube glowing in the darkness.

A figure, X, dressed in white is lying prostrate inside it.

The lights come up. X rises and performs a series of movements, the first Platonic Sequence, evoking the ruler and compass construction of the hexagon. X steps back to admire the construction and then returns to the prone position. The lights dim.

After a slow count of eight the lights come up again and X rises to perform the second Platonic Sequence – the proof of the irrationality of the square root of 2. X notices that the diagonal line from one corner of the cube to the other is the length X just constructed. X paces it out enthusiastically with smaller and smaller steps.

The lights dim and X returns to the prone position.

From offstage-right, we hear the whoosh of a sliding door. A short pause. Then whoosh again. Longer pause.

The whoosh is heard again immediately followed by Y entering stage-right. Y turns and closes the imaginary door. Whoosh. Y is laden down with stuff: a white cubed box, a white bag with shoulder strap and a white cube, slung over the shoulder. Y scans the space and notices a knife sticking into the floor. Y sets the white cube down, sits on it, picks up the knife, examines it and puts it carefully into the bag.

Y stands, picks up the cube seat and moves towards the next space: X's cube. We hear the whoosh as Y opens the imaginary sliding door into X's cube. Y takes one step into X's cube, sees X lying prostrate on the floor, hesitates, and then retreats back closing the door. Whoosh. Pause. X raises their head. Pause. X returns their head to the floor.

Y opens the door. Whoosh. X quickly raises their head. Y immediately recoils and closes the door. Whoosh. X gets to their feet. X and Y stand in their own spaces facing the audience. X prowls round the inside of the cube. Y braces themself and decides to re-enter. With X facing stage-left, Y opens the door, whoosh, steps confidently into the cube and, whoosh, closes the door behind them just as X turns around. They stand facing each other in states of high alert. They start circling each other, like wild animals, sizing each other up.

X moves towards Y.

Y Y.

X freezes and speaks as if for the first time.

X Y.

Y Y.

X One.

Y One.

X One.

Y Y equals one.

X Y equals one.

Y Y equals one.

X Multiply by Y.

Y considers each response carefully.

Y Y squared equals Y.

X Take one.

Y Y squared minus one equals Y minus one.

X Factorise.

Y Y plus one times Y minus one equals Y minus one.

X Cancel.

Y Y plus one equals one.

X Therefore Y equals . . .

Y Zero.

X Therefore Y doesn't exist.

X is perplexed that Y still exists. Y places the cube seat between them. This time Y's responses are immediate.

Identify.

Y Y.

X Quantify.

Y One.

X Y equals one.

Y Y equals one.

X Multiply by Y.

Y Y squared equals Y.

X Take one.

Y Y squared minus one equals Y minus one.

X Factorise.

Y Y plus one times Y minus one equals Y minus one.

X Cancel.

Y Y plus one equals one.

X Therefore Y equals . . .

Y Y equals zero.

X Therefore Y doesn't exist.

Y You mustn't divide by zero.

Y carefully places the white box on top of the cube seat. Y starts to open the box but as X tries to look inside, Y snaps it shut.

Identify.

X X.

Y Quantify.

X One. Unique!

During this exchange we see Y removing something from the white bag.

Y Therefore X equals one.

X X equals one.

Y Multiply by yourself.

X X squared equals X.

Y Differentiate yourself with respect to yourself.

X Two X equals one.

Y Therefore X equals . . .

Y is holding a large tightly folded laundry bag.

X X equals a half. Aghh.

X bends forward in half in agony, at the same time as Y unfolds the laundry bag from an eighth size to a quarter size.

Y Square yourself.

X X equals a quarter. Agghhh.

As X becomes smaller, Y unfolds the laundry bag from a quarter size to a half size.

Y Cube yourself.

X X equals one sixty-fourth. Aggghhhh.

X is now on the floor and as small as they can physically make themself. Y opens the large laundry bag completely, covers X with it, rolls X over so that X is totally inside the bag and, using the bag handles, attempts to lift it. Realising that it is too heavy, Y drops the handles.

Y Raise yourself to the power of . . .

X No . . . No . . . No . . . No . . . No . . . No . . . No . . . No . . .

Y . . . of zero.

X is released from Y's mathematical spell. Arms and legs rupture from the bag. As X escapes and pats themself down, Y folds the bag.

X X equals one. X equals one. X equals one.

Pause.

You can't differentiate with respect to a constant.

Y finishes refolding the laundry bag and places it inside the box.

Identify the coordinates of your origin.

Y There.

Y points to stage-right.

X There?

X looks at the point in mid-air where the end of Y's finger briefly existed.

And the conjectural coordinates when your velocity vector hits zero?

Y Out.

X Out?

Y Out.

X Oh . . . OUT . . .

*X begins performing the first bit of the first Platonic
Sequence. Y watches X for a few moments before . . .*

Y No No No No No No NO! OUT!

*Y points with both arms in the direction stage-left. X
attempts to mimic Y.*

X OUT!

Pause.

Define your terms.

Y A way out! This can't be all there is? A long line of
rooms!

X Room . . . *s*?

Y Yes. Rooms.

*Whoosh. Y opens the imaginary door in the stage-right
side of the cube. X is horrified.*

X You've created a singularity in the side of my cube.

*X looks accusingly at Y and moves towards the open
imaginary door. X stares through it and is stunned by the
prospect. X feels compelled to step forward as Y mimes
gently pushing X from across the cube.*

There exists another room.

Y And another one! And another one! And another one!
They're all the same. It's room, after room, after room.

X An infinite number of rooms!

Y No, no, no. I didn't say infinite. I said it's room, after
room, after room.

X I think I can prove it's infinite.

Y Really? How?

X Induction.

Y Induction?

X Let $P(N)$ be the inductive hypothesis:
'There are N rooms in this line.'
Let $N=1$.
My room exists.
Therefore $P(1)$ is true.
The base step is proved.
Suppose $P(N)$ is true.
I exit to find another room,
Isomorphic to the Nth room.
Therefore $P(N+1)$ is true.
The inductive step is proved.
Now apply the second order axiom of inductive logic
to deduce that
$P(N)$ is true for all N.
Therefore the line of rooms *is* infinite.
Quod erat demonstrandum.

Y Yeeeeees. But if it's infinite then there's no way out.

X There is no OUT.

Y There must be something else . . . this can't be all there
is.

X A manifold needs no ambient space in which to exist.

*A train is heard in the distance. This should seem like a
real train passing underneath the theatre, the sound
subtly intruding on the quiet of the performance space.
This will be a recurrent sound that we will hear at key
points in the play.*
X steps back into the new room.

It's just room . . .

*Whoosh as X opens another imaginary door, exits stage-
right and steps into another new room.*

X (*off*) . . . after room.

A fainter whoosh.

. . . after room.

We hear X's voice trailing off into the distance repeating: whoosh . . . 'after room'.

Y sits down on the cube seat. Y takes the white bag and from it produces an orange. Y begins to peel it. Orange zest explodes into the atmosphere.

Enticed by the smell, X re-enters stage-right. Smell is a new, physical experience for X. Seeking its origin X re-enters the cube and sees the orange for the first time. X is transfixed.

X It's . . . It's . . . It's an X squared plus Y squared plus Z squared equals R squared.

Y It's an orange.

X It's a norange sphere.

Y Yes, a norange sphere.

X The norange sphere is establishing a functional relationship with my nose.

X is staggered as Y eats a piece of orange.

A segment of the norange sphere has become embedded as a subset of Y.

Y offers a piece of orange to X. X hesitantly takes it. They bite down and chew the fruit together. It's X's first experience of taste. X is overwhelmed.

X has been functionally injected.

Y offers X a second larger piece of orange which is readily accepted. Together they eat the whole orange. Y takes another orange out of the bag but before Y can peel it, X stops Y.

X Redefine the origin of the orange sphere using the following equation: X minus fourteen squared plus Y minus fourteen squared plus Z plus two squared equals R squared.

Y is unsure what to do, so X repeats the instructions.

X minus fourteen squared.

Y holds the orange carefully in outstretched hands and allows it to lead Y along the stage-left side of X's cube to the front corner.

Plus Y minus fourteen squared.

Y moves the orange along the front of X's cube ending above X's waiting hand.

Plus Z plus two squared equals R squared.

Y drops the orange into X's hand. Thrilled, X looks lovingly at the orange.

One! One!

X looks for somewhere special to place it.
    Y picks up the cube seat and offers it to X as the perfect podium on which to position the precious orange. X carefully sets the orange down, stands back and admires the physicalisation of the number 1.

One!

Pause.

Another 'one'?

Y looks into and counts the remaining oranges in the bag.

Y One, two, three, lots. Yes.

Y gets out another orange and begins to move it through the space as before but X stops Y.

X The origin of the orange sphere should follow the following path defined by the equation Z equals Y minus G divided by two Y squared where G is the acceleration towards the floor.

Y throws the orange underarm in a parabola into the outstretched hand of X, who catches it.

One!

X places it lovingly next to the first orange.

One. Two. One! Two. One. Two! One! Two! ONETWO!

Pause.

Another 'one'?

Y counts the remaining oranges in the bag.

Y One, two, three, lots. Yes.

Y produces another orange and waits.

X The origin of the orange sphere should follow the following path defined by the equation X equals *e* to the minus T cos T, Y equals *e* to the minus T sine T, Z equals one over T where T is a parameter running from one to infinity.

Y That's not physically possible.

X It's a well defined function.

Y goes to put the orange back in the bag. X panics.

An *approximation* would be acceptable.

Pause.

The orange sphere?

Pause.

Conjecturally?

Y relents and attempts to spiral the orange infinitely down into X's out stretched hand.

X An impressively small epsilon . . . One!

Y Humph.

X places the third orange next to the first two.

X One. Two. Three! Another 'one'!

Y counts the remaining oranges in the bag.

Y One, two, three . . . no.

X Oh.

A deflated X prowls the cube, deep in thought.

A multiplication inverter?

Pause.

An integer divider?

Pause.

A disconnected space creator?

Y A knife?

Y produces a knife as if from nowhere. It is the knife Y picked up earlier.

X Knife?

X takes the knife, runs a finger along the blade and understands what it can do.

Knife!

X cuts one of the oranges in half.

One, two, three, four!

Y What are you doing?

X Creating physical number. I've cut this orange sphere in half with your 'knife' and now I have one, two, three, four pieces of orange sphere.

Pause.

If I cut this half in half again with your 'knife' then I will have one, two, three, four, five pieces of orange sphere.

Pause. Stands and looks from the oranges to stage-right.

If I apply the same inductive reasoning that proved that the line of rooms is infinite, then after infinitely many cuts I will have infinitely many pieces of orange sphere.

X prepares to start cutting.

Y I'm afraid that's not physically possible.

X It may take me some time.

Y It would take you more than some time.

X How much time?

Y It would take you an infinite amount of time!

X That is a long time . . .

Y Especially towards the end. You'd have to spend the rest of time cutting that orange sphere.

Pause.

X But . . . what if I cut one orange sphere in half in eight seconds? And then I cut that half in half in four seconds. Then I cut the quarter in half in two seconds. Then I cut the eighth in half in one second. Then the sixteenth in half in half a second. The thirty-twoth in half in a quarter of a second . . .

X speaks faster and faster to mimic the increasing speed of the cutting.

Y (*interrupting*) It's alright. I can see the pattern.

X Yes! So after infinitely many cuts I will have infinitely many pieces of orange sphere. But it won't take me an infinite amount of time! It will take me the sum from N equals minus three to infinity of two to the minus N . . . I can create infinity in sixteen seconds.

Y That's a lovely idea.

X Thank you!

Y But I'm afraid you won't be able to make it happen in reality.

X Why not?

Y You'd have to cut faster and faster and faster, until eventually you'd be cutting so fast that you'd hit the speed of light, and nothing can go faster than the speed of light. Speed is finite.

X No, I can do it. I can.

Y And space too. You can't just keep on dividing space. At some point you'd hit the atom! Split that and you'd blow the whole place sky high.

X No, it would work. The maths makes perfect sense.

X places the knife down defiantly.

Y Do it then.

X Really?

Y Do it then. I'll time you.

Y produces a chess timer from the bag.

Cut that orange into infinitely many pieces . . . in sixteen seconds.

X I will.

Y I'm going to enjoy this. Infinity created before my very eyes. Your hands moving faster than the speed of light. My knife so sharp that it'll be capable of slicing an atom. GOSH!
I'd better put on my safety goggles.

Y takes a pair of safety goggles from the bag and puts them on.

There now, perfectly safe. Are you ready?

X Yes.

Y Pick up the knife.

X picks up the knife.

Sixteen seconds to create infinity. Ready.

*X prepares like a runner at the start of a race. X false
starts.*

Steady.

X hesitates as Y shouts:

Cut.

Y One.

*X runs to the orange and cuts it quickly in half and then
stands confidently admiring the first cut.*

Two.
    Three.
    Four.
    Five.
    Six.
    Seven.
    Eight.

X quickly cuts the half in half and stands again.

Nine.
    Ten.
    Eleven.
    Twelve.

X cuts the quarter in half.

Thirteen.
    Fourteen.

X cuts the eighth in half.

Fifteen.

As Y continues to count, X's confidence evaporates as X cuts more frantically.

Sixteen.

Y STOP CUTTING!

X tries to squeeze in an extra cut.

PUT DOWN THE KNIFE!

X puts down the knife and looks bemused.

STEP AWAY FROM THE ORANGE!

X's hands are covered in orange juice.

So you have created infinity!

X I didn't realise equations were so sticky.

Y But you have created infinity?

X Well . . .

Y Nearly?

X Errrr . . .

Y How far did you get on your way to creating infinity?

X One, two, three, four, five, six, seven.

Y SEVEN?!?

X I just need more practice.

Y Especially towards the end.

Y goes to the bag and takes out a damp sponge contained in a plastic bag, tidies the cut oranges away and wipes the top of the cube.

X But I proved it. I can infinitely divide numbers and add them up. So why can't I do that to the orange sphere?

Y Because the orange is real but infinity isn't. Infinity only exists in the imagination. Not in real true life.

X It does.

Y It doesn't.

X It does.

Y It doesn't.

X It does.

Y It doesn't.

X Does.

Y Doesn't.

X Does.

Y Doesn't.

X So, so, so . . . if infinity doesn't exist . . . what's the biggest number then?

Y Oh! The biggest number is seventy-three million and twelve.

X What about seventy-three million and thirteen?

Y Gosh! I was very close.

Pause.

X NO! Numbers don't run out. Infinity does exist.

Very upset, X heads to the back of the cube and stands facing upstage in a sulk.

Y Come and sit down, we need to have a little chat.

Y places the cube seat in the centre of the room, sits on one side and invites X to sit too. X hesitates and then awkwardly accepts Y's invitation. They are conscious of the close proximity of their bodies.

Numbers are just in your mind.

X But I just physicalised . . .

Y In your life, you only have a finite amount of time, so there will only be a finite number of numbers you will ever need. So, for you, and for me, there will be a biggest number. One day, you will think of a number, and after that, you will never think of a bigger one, because you'll die before you get the chance.

X Die?

Y Yes, die. Your cells won't reproduce any more. The telomeres in your chromosomes will get shorter and shorter until they're all used up. The last number. No more plus one. The end of the line.

X But a line has the potential to go on for ever.

Y Ah! Potential! What a marvellous idea! Teeming with possibilities. And for ever . . . I think that's my favourite idea of all. 'They all lived happily ever after.' No. You can prove to me that infinity is a lovely consistent idea, but I'm afraid you'll never be able to SHOW it to me.

Y stands up and paces the cube, pointing out each corner.

Take this space. It's all finite lines. It starts here.

Y points to the stage-left front corner.

And it ends over there.

Y points to the stage-right front corner.

You're born there.

Y points to the stage-right back corner.

And you die there.

Y points to the stage-left back corner, where Y places the cube seat.

And this line of rooms too, it starts somewhere and it ends somewhere. It's all finite.

Pause.

X But there is infinity in my cube.

Y Really?

X (*sotto*) You promise you won't tell anyone?

Y (*sotto*) Okay . . .

X Just before you arrived, I discovered infinity hiding in this cube.

Y Where?

X From there to there.

X points from the stage-right back corner of the cube to the corner diagonally opposite.

Y That's a finite length!

X Not so! If the side of my cube has length Y . . .

X paces out the downstage side of the cube.

Y That's me!

X Yes . . . And the diagonal across the room has length X . . .

X points down the diagonal across the cube.

Y That's you!

X Yes . . . Then X is bigger than Y by a factor of the square root of two. And that's an irrational relationship.

Y What do you mean, that's an irrational relationship? Who are you calling irrational? I'm not irrational!

I'm extremely rational, thank you very much. What a ridiculous thing to say: 'That's an irrational relationship'?!?

X No, no, no . . .

X gestures at the two lengths.

X and Y are incompatible.

Y Incompatible! What do you mean, we're incompatible? How can you possibly know if we're incompatible? We've literally just met.

X No, no, no . . . If you write the square root of two as a decimal number, then it goes on to infinity, never repeating itself.

Y So you won't be able to show it to me!

X No, I can! . . . With my steps! . . . It starts . . . one . . .

X takes two large steps down the diagonal.

Point four . . .

X takes a smaller step.

One . . .

X takes a further tiny step repeating the action from the second Platonic Sequence seen in the opening of the play.

Y Do you mind if I make myself comfortable while you do that?

Y takes the cube seat, places it in the centre of the space and sits on it.

We may be here for a while. No need to rush . . . we've got plenty of time.

From the bag Y produces a sandwich wrapped in greaseproof paper, which Y eats during the following exchange.

X Four . . . two . . .

Y Very good. Only an infinity to go. You do know, don't you, that the rule of decimal numbers means that each step you take must be a magnitude of ten times smaller than the one before.

X I know, I know.

X is now on the floor in a plank position trying to measure out the space with the tips of their fingernails.

Y I know you know, but just so you know, your next step should be the size of a pollen grain . . .

Y takes a bite from the sandwich.

Measured that?

X (*strained*) Yes . . . one . . .

Y And the next step should be the size of a bacterium . . .

Y takes another bite.

Measured that?

X (*increasingly strained*) Yes . . . three . . .

Y Your next step should be the size of a virus on a bacterium . . .

Y takes another bite.

Measured that?

X (*strained to breaking point*) Well . . .

Y Nearly?

X Aaaaahhhm!

Y Really?

X NO.

X collapses to the floor in great distress.

Y Never mind. You nearly measured to seven decimal places.

X I couldn't do it.

Y Don't be upset. You tried to do something else that isn't physically possible.

X I don't understand. Why couldn't I do it?

Y Because there's a limit.

X There's no limit, I could have kept measuring if only I'd had a . . .

Y No, no, no. There's a physical limit. Even if you'd got to the thirty-fourth decimal place you couldn't go any further because of the Planck constant.

X The Planck constant?

Y Yes, the Planck constant. The smallest ruler that could possibly exist. *Don't* try to measure any smaller than the Planck constant because you'd make a mini black hole . . . which would be awful.

X So I'm not going to be able to show it to you?

Y No. Never mind. Have a sandwich.

X accepts the sandwich. It's a right-angled isosceles triangle with root 2 down the diagonal. X traces out the line, traumatised.

X Root two!

X swallows the sandwich whole, trying to get rid of the failure to measure root 2. It's Marmite. X doesn't like it.
Pause.
The sound of the offstage train is heard.

Y Well, as much fun as this has all been, we are going to have to admit, aren't we, that that infinity is also in the mind.

X I know something that really is infinite.

Y Really?

X Time! You said we have plenty of time! Time goes on for ever! Time is infinite.

Y becomes very serious.

Y No, no, no, no, no. Time doesn't go on for ever. Time has an end. That end even has a name. The heat death of the universe.

Y stands, gradually moves to the front of the cube, and speaks with extreme gravity.

The second law of thermodynamics states that if the universe lasts for a sufficient time, it will asymptotically approach a position where it has diminished to a state of no thermodynamic free energy, and therefore can no longer sustain processes that consume energy, including computation and life.

Pause.

X Shit.

Y Oblivion.

X That infinity looks a bit bleak . . . especially towards the end.

Pause.

Well, what about this line of rooms? It's infinite. I proved it to you.

Y That proof of yours, you do know that there was something not quite right about it, don't you?

X Not quite right!

Y Just because I've been through lots of rooms that don't contain OUT doesn't mean that there isn't OUT in the next one. Here, take this coin.

Y produces a coin from the bag.

X You've projected the sphere into two dimensions.

Y It's got two different sides, a head side and a tail side. When I toss it, like this, it is equally likely to land tails side as land heads side.

Y performs a simple coin toss.

Heads. Every time I enter a new room it could contain an OUT. Just as every time I toss this coin it could land tails side. It doesn't matter how many times I've tossed it and got heads, that doesn't mean that the next time I toss it I won't get tails. Your 'proof' would imply that this coin always lands heads. Even if I'd tossed ten heads in a row there's still a fifty-fifty chance I'll get a tails on the next throw.

X How many rooms have you been through?

Y Seventy-three million and twelve.

X And it always lands heads?

Y Yes.

X An unbroken sequence of seventy-three million and twelve heads? Are you sure your equation hasn't lost its tail? Perhaps it's the same on both sides.

Pause.

Y The same? The same! Your proof assumes all the rooms are the same.

X Of course. You said. They're all the same. Room after room after room.

Y But they're not all the same.

X Really?

Y The rooms look the same, but what they contain is different. How do you think I got those oranges?

X I thought you blew up the singularity at zero?

Y No. I found them in some of the rooms I went through. There are other fruits too. Pomegranates. Kumquats. Lemons.

Y produces a lemon from the bag. X takes the lemon.

X X squared divided by A squared plus Y squared divided by B squared plus Z squared divided by C squared equals R squared.

Handles it lovingly.

An ellipsoid.

Y Some celluloid. A vuvuzela. A bagel.

Y produces a bagel. X places the lemon on the cube and takes the bagel.

X R one minus the square root of X squared plus Y squared all squared plus Z squared equals R two squared.

Smells the bagel.

A genus one surface!

Y A slinky. A bicycle. A book.

Y produces a book from the bag.

X A cuboid.

X places the bagel next to the lemon and takes the book. Its pages flap open.

Did you see that? The manifold has decomposed itself into a foliatation of parallel submanifolds with a singularity running right through the middle!

Y Now see here, this room wasn't empty.

X Well it was before you brought all your things in here.

Y No. You were in it. This is the first time I've been into a room and found another person. Have you ever met another person?

X Person?

Y Like me.

X Another variable! No. I thought I was unique.

Y Fascinating! If a room can contain you, it can contain OUT. What's behind that door there?

Y points stage-left.

X I don't know. I've never looked.

Y You've never looked! Of course you've never looked. Why would you have ever looked?

X I thought this was all there was.

Y So OUT might be on the other side of that door.

Y moves to the door, hesitates, then opens it. Whoosh. Y steps through the imaginary door into the next cube, then closes the door. Whoosh.
   Y has disappeared from X's view. On the other side of the door is yet another room. It isn't OUT. Y is deflated. Pause.
   On the floor nearby is a ball of string. Y picks it up and examines it.

X Are you out?

X is confused by Y's leaving and experiences feelings of abandonment. Another new emotion. Pause.
X re-examines the items Y has left until Y angrily slides open the imaginary door. Whoosh.
   Y re-enters the cube and closes the imaginary door. Whoosh. X is glad to see Y return.

X So how did the coin land?

Y (*through clenched teeth*) Heads again.

X What?

Y (*barely audible*) Heads again.

X A hexagon?

Y (*shouted*) I said heads again!

X Oh, heads again! So it is infinite.

Y thrusts the big ball of string at X.

What's that?

Y String. I found it in the next room. And this ball of string proves that your proof doesn't work.

Y offers the string to X, who is moving around trying to avoid it.

If I can find a ball of string in that room, then the next room might contain OUT. STOP moving around! You can't actually prove to me that it's infinite, can you?

X No.

This truth is devastating to X.

Y Exactly. Put those things down. Take this. In *both* hands. Very carefully.

X puts the other items down and takes the string as ordered.

This line of rooms isn't infinite. Infinity doesn't exist.

X It d—

Y This line of rooms has a last room. It has an exit, it has an OUT and I'm going to find it. And do you know what? I am going to prove it to you.

X How?

Y With string!

X String?

Y Stand there. Don't move.

X does exactly as told and stands still as Y packs the items back in the bag and puts it on. Y then picks up the cube box, followed finally by picking up the cube seat, thus replicating our first image of Y.

I am going to take the end of this string and I am going to walk through that door into the next room, and then into the next room, and into the room after that, if necessary, and I am going to keep on walking in a straight line through room after room after room until I find OUT. And when I find it I am going to pull on this string, three times, like this –

Y pulls the end of the string three times. X's hands get yanked three times.

– to PROVE to YOU that I HAVE found it.

X How long is your ball of string?

Y It's as long as I need it to be.

X You're going to need an infinite ball of string.

Y We'll see . . .

Y slides open the stage-left imaginary door and walks through. Whoosh.
Y exits completely stage-left. Watching Y leave, X stands with the large ball of string which dances in X's hands as if playing out in the direction of Y's exit stage-left.
After some time Y re-enters stage-right still holding the end of the string and walks slowly towards X, who is still facing stage-left.
Y enters the cube, sees X and stops.

Y pulls on the string three times. X's hands get yanked forward three times. Surprised, X pulls back on the string once in response and Y's hands get yanked back. X suddenly becomes aware of Y's presence.

X Who are you?

Y (*Y's voice has aged*) Y.

X How many are there of you?

Y One.

X No!

Y Y equals one.

X No! Y is *that* way.

X points stage-left.

Y Y *is* that way.

X Y said they were going to go in a straight line that way.

Y I did go in a straight line that way.

It dawns on X that this really is Y.

X Then how did you come from there?

X indicates stage-right.
Y wearily puts down the cube seat, the bag and the box. X and Y both regard the two ends of the string they hold before physically joining them together. They both imagine Y's journey, their gaze traversing a large circle in the air. They arrive at the same conclusion.

Y It's all joined up!

X It's a loop!

Y There is no OUT.

X It's not infinite.

X *and* **Y** We were both wrong.

They both throw their ends of the string into the wings.

Y We were both wrong. Oh dear!
Oh dear, dear, dear.
How very disappointing.
I'm so very tired, you see . . .

X notices that Y needs to sit, and quickly slides the cube seat over so that it stops directly behind Y. Without hesitation Y sits.

. . . I've come a very long way,
I've collected all this . . . stuff,
I've carried . . . all this . . .
I kept everything I came across,
EVERYTHING! . . .
and for what?
Why?
What has it brought me?
What's the use of it all?
What does it all mean?

X You've got everything!

Y I've got nothing . . . NOTHING! I'm just stuck in this line of rooms.

Pause.

X You've got *me*.

Y Yes.

Pause.

X X squared plus Y squared equals one.

X traces out a circle in the air.

Y Codependent variables.

X When X was one, Y was zero.

Y And when Y was one, X was zero. But now . . .

X X equals Y.
X squared plus Y squared equals one.
X and Y are both half the square root of two.

*X has moved to the downstage-right corner of the cube.
X realises that this number is half the length across the
diagonal of the room.*

Y That's very nice, dear.

*X reaches out to the centre of the room and Y stretches
out as if to join hands with X. X suddenly notices the
aged skin on Y's hand and recoils.*

X Your hands? Your face?

Y I was walking for such a long time.

X But you only just left.

Y I've been away for many, many years and now . . . I'm
approaching my singularity.

X Singularity?
A singularity is a point which is not defined, which
doesn't exist.

X realises the huge significance of this.

No . . . no . . . we've only just started.

Y X times Y equals one.

X You inverted the circle. Is that a way OUT?

Y No.
But as Y tends towards zero . . .

X X tends towards infinity. So I get my infinity back!

Y Yes, well I suppose you do, but I'm afraid you lose me in
the process.

X I don't want that sort of infinity.

Y I'm afraid that's reality, dear. There's nothing much we can do about reality.

X I can resolve your singularity.

Y No, I'm afraid you can't. But perhaps you can tell me about *your* OUT.

X My OUT!

X begins to describe their OUT. Y doesn't resist.

A place where numbers go on for ever . . . where parallel lines meet . . . where I can create new shapes, new numbers.

Y Like the square root of two?

X Yes. Like the square root of two.

X begins to perform the second Platonic Sequence proving the irrationality of the square root of 2.

If you construct a square with sides of length Y, then the diagonal X is never compatible with Y, the sequence of diminishing squares never terminates . . .

Y Can you make a new number . . . ?

X A new number?

Y Yes.

X A new number. Create a new number . . .

Y Do you mind if I just close my eyes for a moment, dear, while you do that?

Y closes their eyes and their head flops gently back as Y falls asleep.

X One, two, three, four . . . nothing new there . . .
Minus one, minus two, minus three, minus four . . .

X shakes their head.

A half, a quarter, an eighth, a sixteenth, one thirty-twoth . . .

*As X becomes more desperate Y, unseen by X, silently flops
their head forward and collapses their upper body, dead.*

Nothing there either . . .
Square root of two . . .

X traces out the square and its diagonal.

The circumference of a circle . . .

*X traces out a circle but abandons it and in desperation
turns to Y.*

I've got nothing, Y.
There's nothing else. Y?

X notices that Y is not responding.

Y!
What's happened to you? Why aren't you listening to
me? Y equals one . . .

*X goes to touch Y but doesn't, instead letting their hand
hover only inches above Y's back in a moment of
suspension.*

Y equals one . . .
Y squared equals Y.
Y squared minus one equals Y minus one.
Y plus one times Y minus one equals Y minus one. Y plus
one equals one.
Y equals zero. Nothing.

X realises that Y has hit their singularity: Y is dead.
*X stands at the front of the cube, eyes closed frozen in
grief.*
The lights begin to fade as if the play has finished.
In near darkness X suddenly speaks.

There must be something else!

*The lights snap on, washing the stage with what would
be the lighting state for the final curtain call.*

This can't be all there is?

Y looks up, puzzled, then returns to their death position.

The square root of nothing . . . is nothing.
   The square root of one . . . is one . . . and minus one. The square root of minus one.
   There isn't a number which when you square it gives you minus one.

Y Sshhh.

Pause.

X Take the square root of minus one. Imagine it.
   X squared equals minus one.
   Plus times plus is plus; minus times minus is plus; so the square root of minus one would be a new number.

Y looks up again, perplexed at what X is doing. The play is meant to have ended.

Y (*through gritted teeth*) Stop!

X is double checking their calculations.

X It's self-consistent, non-contradictory.

Y (*sotto*) Shut up.

X The square root of minus one. It's a new number.
   A new world.

X looks up and points to the ceiling of the cube.

A new direction . . .
   I did it . . .
   There's an OUT.
   I've done it, Y. I've created OUT!

Y raises their head.

Act Two

THE THEATRICAL DIMENSION

Y Stop it!

X turns to see Y returning to their dead position.

X You're alive?! It worked.

Y sits up and speaks through gritted teeth.

Y No. I'm dead.

Flops dead.

X You're alive!

Y No. I'm dead.

X You're not dead!

Y rears up angrily.

Y I am dead. I died.

Dramatically flops dead.

X And then I saved you with my new number.

Y (*through gritted teeth*) You didn't save me. It's not possible to save someone when they are already dead.

X Then how come you're alive?

Y I'm not alive. I'm dead.

X I don't understand.

Y rears up.

Y (*emphatically*) It's very important that the play ends with my death. It is symbolic of the nothing that awaits us all at the heat death of the universe.

Dramatically flops dead again.

X So you're dead.

Y Yes. I'm dead!

X But you're alive.

Y Of course I'm alive.

X I still don't understand. How can you be dead and alive at the same time?

Y springs from the seat and explodes.

Y I was ACTING!!!

Y's whole demeanour has changed.

Everyone here was complicit in the act of imagining that I was dead.
We were telling a story.

Pause.

X Have we finished the story?

Y Yes . . .
Well, no . . .
We're still here.

X So you're not going to die.

Y No . . .
Well, yes . . .
But not now . . .
I hope . . .

X That was a theoretical death?

Y It was a theatrical death!

X None of this is real?

Y No . . .
Well, yes . . .

I am real . . .
I hope.

X And I am real . . . aren't I?

Y I don't know.

X But you really are Y.

Y No.
Y is a character in the story. I am an actor called Victoria who was pretending to be Y.

X Pretending to be Y?

Y Y equals Victoria.

X Y equals Victoria? Victoria equals one.
Multiply both sides by Victoria . . . Victoria squared equals Victoria.
Victoria squared minus one equals Victoria minus one.
Factorise.
Yes, I suppose that still works. Victoria.
So X equals?

Y X equals X.

X Oh.

X looks crestfallen.

Y Some of this is real.
We're in a real space. With all these lovely real people.

X People?

Y Yes. We have an audience.

From this point until the end of the play the character of Y is aware of, and frequently acknowledges, the audience.

X More variables!

Y They are why we are here.

X What are they looking at?

Y They're looking at you, and me, and our imaginary world.

X Where nothing is real.

Y Yes! . . .
Well, no!
Yes . . . this bag is real, these trousers are real trousers.

X And my room. That's real.

Y No! . . .
Your room isn't real. It hasn't even got real doors.

Y mimes opening and closing the stage-right door of the cube making the whoosh sound.

The cube is real but it only exists because it's part of a theatre set that we had specially built for this show.

X But if I go through that imaginary door, there are more cubes that way . . .

X points off in the wings stage-right.

And more cubes that way . . .

X points off in the wings stage-left.

Y No! There's just this one cube.
Actually that's not true. There's also the model that our designer insisted we had made, at vast expense, so that we could see what the *cube* was going to look like before we built it for real.

During this, Y extracts a small model cube from the white cube box. X moves the cube seat downstage so that Y can place the model cube on it.

X That's my cube. Is it real?

Y Yes . . .
Well, no . . .

It's a model of your cube . . .
And here's a model of you.

Y produces from the white box two small cut-out figures of X and Y on sticks.

And here's a model of me coming into the cube.

Y performs a section from the opening scene of the play as if in a tiny puppet theatre, including mimicked voices of X and Y.

Y (*as X*) One.
(*As Y.*) Y equals one.
(*As X.*) Multiply by Y.
(*As Y.*) Y squared equals Y.

X Take one.

Y (as Y) Y squared minus one equals Y minus one.

X *and* Y (*as X*) Factorise.

X Do I really sound like that?

Y (*as X*) Yes you do rather, actually, sound quite a lot like that . . . actually.

Y reverts back to their own voice.

So you see the model helped to make the story.

X And the story is real.

Y No! . . .
Well, yes.
It is a real story, but the story's not REAL. We made it up.

X I made myself up?!
That's a strange loop.

Y Yes . . .
Well, no . . . You're in the script.

Y takes the script from the white box and finds the last page.

Which ends with Y's death.

X Script?

Y A script is an executable section of code that manipulates variables to tell a story.

X Am I in the script now?

Y No.
We're not even meant to be here any more . . .
The play has finished . . .
We're off script.

Pause.

X So what are we doing now?

Y I don't know . . .

A very long pause as they stand doing nothing, looking out at the audience.

I'm not really sure what's happening . . .
In fact I'm not sure I know what is real any more . . .

X The script is real.

Y Yes, but it's just bits of paper.

X It's got the code written on it.

Y The code isn't really the script.
The script is the story and the ideas. It's the ideas that are really real.
And I'm beginning to feel that perhaps the mathematics in the story is arguably the most real thing of all.

X So infinity is real.

Y Yes . . .

X I knew it!

Y Well, no.
  Well, yes . . . it's a real idea . . .
  But no . . . it's not real . . . you can't show it to me.

X Then what is real?!

Y (*grappling for the truth*) What is real?
  What is *real*?
  The quest for meaning. That's real!
  The desire to subvert the finite. THAT'S real.
  The NEED for happy ever after.

X You mean the search for OUT.

Y Yes. I suppose I do.

X Because you know when you were dying and you were
going to leave me on my own in here with my infinity and
I thought I was going to miss you rather a lot, I had just
found you OUT.

Y Really?

X I created a new number, you encouraged me and
I imagined it.
  And there it was. A new direction.

  X proudly points to the ceiling of the cube.

Y How is that OUT?

X I thought you could have climbed out there.

Y How could I climb up there?

  *Y exits stage-right without the need to open or close
  imaginary doors.*

X I hadn't got to that bit.

  X is unaware of now being alone on stage.

How could you climb up there? Good question. Perhaps
your model will help to find a solution.

X moves to the small model and starts improvising a scene, mimicking the voices of X and Y.

(*As Y.*) How could I climb up there?

(*As X.*) I don't know . . . I'd only got as far as creating a new number . . . I hadn't worked out exactly how to do it yet . . .

(*As Y.*) Perhaps there is something in the box that might help . . .

(*As X.*) In the box?

(*As Y.*) Yes . . .

(*As X.*) In the box. In the box!

X goes to the real white box and, after rummaging, pulls out a mini-stepladder. X returns excitedly to the model theatre.

Look what I've found! You could climb out on this!

X places the mini-stepladder in the cube, picks up the model figures and continues the improvised story. Unseen by X, Y enters stage-right with a real stepladder and places it in the centre of the cube with the steps facing squarely downstage. The stepladder has a black cloth attached to it which masks the upstage side of the steps.

(*As Y.*) Oh you are so very clever, X.

(*As X.*) Thank you.

(*As Y.*) You really are so very, very . . .

Y I don't sound like that.

X Well you do sound a bit like that . . . sometimes.

Y climbs up the real stepladder, and proceeds to climb down the other side, conspiratorially putting their fingers to their lips. Y's descent is masked by the black cloth attached to the ladder. Y exits through the back of the cube.

X improvises the scene where Y climbs the ladder and gets OUT.

43

I thought you could climb up on these. I created a new direction with my new number and I thought that you'd be able to climb up and OUT.

Pause.

I think that this time I would come too because maybe I'd find my infinity. And then we could both be OUT.

Y (*off*) Come on then.

X turns and sees the real ladder.

X It's real.
It's really real! I'm coming, Y!

X turns the ladder ninety degrees so the steps now face squarely stage-left. X starts to climb the ladder but on the first rung is suddenly gripped by fear and starts shaking, unable to climb higher.

X I am oscillating.

Y (*off*) Oscillating?

X It's very high.

Y (*off*) Ah! That's fear. Very real and entirely natural, given the potential risk.

X Risk?

Y (*off*) Of course. What you are attempting to do is *incredibly* dangerous.

X Really? I'm stuck. I can't move.

Y (*off*) That's paralysis. Fear in its most acute state. What you need is a good dose of induction.

X Induction?

Y (*off*) Let P(N) be the statement you can climb N rungs of the ladder. You've got on to the first rung so that's the base step proved.

X Right.

Y (*off*) And when you're on the Nth rung it's a simple step up to the N plus oneth rung.

X Simple for you maybe . . .

X climbs with exaggerated caution to the second rung, terrified.

Oh, that was quite simple.

Y (*off*) Good. That's the inductive step. Now apply the second order axiom for inductive logic and you'll find you can climb as high as you like.

X Really?
Induction.

X starts freely climbing the ladder.

It's working.
I'm coming, Y . . .
To infinity.

As X reaches the top of the ladder, Y returns stage-right carrying a prop trap door that Y is hiding inside. This is a cardboard square with a door. Black cloth is attached to the edge draped to the floor. Y's head pops through the door which when open reveals the words TRAP DOOR *written on the inside.*

Y Keep climbing.

X looks down and sees Y climbing out of the prop trap door as if through the floor of the stage.

X What the factorial!
How the factorial?
What the factorial is that?

The trap door is now lying flat on the stage with Y standing in the square hole left by the door, which is

45

hinged up, with the words TRAP DOOR *clearly visible to the audience. Y steps out.*

Y It's a trick.

X A trick?

Y Yes, a trick. It's a trap door.

X A trap door?

Y Yes, a trap door. It's a trick trap door.

X A trick trap door?

Y Yes, a trick trap door.

X I don't understand. You went up there. How did you come up from down there?

Y Theatrically. If you step inside I'll show you.

X Is it safe?

Y (*to the audience*) 'Is it safe?'!

X stands in the trap door, and Y pulls it and the black cloth up so that X disappears from view.

Y Are you alright?

X's voice sounds as if in a cavernous room.

X It's very dark down here.
Ahhh . . . octagonal variables!

Y (*to audience*) Spiders.

X Urghh . . . Sticky network graph!

Y (*to audience*) Cobwebs.

X I don't like it. Can I come out now?

Y drops the curtained trap door to X's feet. X steps out.

Y You see? It's a theatrical trick.

X I still don't understand . . .

Y I didn't really climb through a hole in the ceiling.

X Really?

Y No. I climbed up the ladder . . . here . . . and then climbed down the back of the ladder . . . here . . . And disappeared behind this black cloth . . . here . . .

X Really?

Y Then I just walked off the stage.

X Really . . . OUT?

Y No, I just walked off into the wings.

X Wings? The space can fly?

Y No. It's a theatrical term. I just walked off the set into the wings.

Points stage-left.

And squirrelled round behind the back of the set.

X Round the back?

Y Yes, there's a way round. Go and see for yourself. It's quite safe.

X cautiously follows Y's instructions.

Step off the set . . . good.
Now walk across the stage. Past the legs . . .

X hesitates.

Don't ask . . .

X exits stage-left.

Into the wings . . . That's it . . . Mind the lanterns.

X (*off*) Ow!

Y And the speaker!
Now walk down the treads and turn upstage . . .
Left . . . turn left . . . That's right . . .

X (*off*) Right?!

Y Not right! Left . . . that's right.

X (*off*) But . . .

Y (*gesturing clearly*) That way . . . go that way . . . That's it . . .
And now you'll find you can just squirrel yourself round behind the blacks.

X (*off*) Squirrel?

Y You'll see.

X (*off*) I can't see. There's nothing, just blackness.

Y You'll be alright. Keep going.

X (*off*) Ahhh! There's another variable.

Y That's just the stage manager. Don't worry about them.

X (*off*) Oh . . . there's a way around!
So I do the squirrel round the back here?

Y Yes.

X (*off*) I can see a light. There's something on the other side!

Y Yes.

X (*off*) Ohhh! It's symmetrical. It's looped.

Y Good! So now you're standing stage-right in the wings. If you come up the treads, and across . . .

X (*off*) I can see Y!

Y Yes, you can . . . Now walk towards me across the stage and on to the set and you'll find you've made an entrance.

X re-enters stage-right.

X So . . . you didn't go all the way round the universe . . .
You just did the squirrel round the back . . .
What happens if I invert the path?

X disappears stage-right to investigate.

Y Yes, you can also go the other way.
(*To the audience.*) I think X is going to make another
entrance. Shall we give them a little round of applause
when they do?

X confidently enters from stage-left, backwards.

X Oh, it is invertible . . .

*On hearing the applause X turns to face the audience. At
first X is perplexed, but then rather enjoys the feeling
until the applause ends.*

X Again?

X swiftly exits stage-left.

Y I think they rather enjoyed that. Shall we indulge them
one more time?

*X enters and receives the applause with very theatrical
bowing until the applause ends.*

X Again!

Y No . . . don't milk it.

X Milk it?

Pause.

Y Never mind. So that's how we suggested that the space
was doubly looped. Theatrically.

X So my new number didn't make you an OUT.

Y I'm afraid not.
Now we'd just be stuck on the surface of a gigantic
orange . . . looped in two different directions . . . looped
that way . . .

Y indicates the first large circle running from stage-left to stage-right.

. . . and that way.

Y indicates a second circle perpendicular to the first circle going out through the roof and coming up through the floor.
We hear the offstage train rumble.

X Ohhh! . . . or we could be on the genus one surface you brought in . . . The R one minus the square root of X squared plus Y squared all squared plus Z squared equals R two squared.

Y (*quietly*) The bagel.

X What did you call it?

Y The bagel.

X Exactly . . . Where is it . . . the bagel?

Y I ate it.

X You ate it!

Y I was hungry.

X You ate the universe!

Y It was just a prop.

X Quite an important prop, actually.

Pause.

What's a prop?

Y A prop . . .
Props are part of the physical geometry that help us to make theatre.

X Are they real?

Y takes out a child's exercise book from the bag.

Y Sometimes but not always. In theatre we suggest, and by suggesting we can make something seem to be real.

Y opens the book and uses it to suggest a bird that momentarily takes flight. X is transfixed but can't understand where the bird goes when Y closes the book.

That's the power of the imagination. In theatre anything is possible.

The theatre is the set of all possible self-consistent structures.

Y puts the exercise book in the box and takes out the cuboid book we saw earlier.

X So am I a prop?
Am I what I think I am?

Y opens the book and starts to read.

Y
'Can this cockpit hold
The vasty fields of France? Or may we cram
Within this wooden . . . *cube* . . . the very casques
That did affright the air at Agincourt?
O, pardon! Since a crooked figure may
Attest in little place a million;
And let us, ciphers to this great accompt,
On your imaginary forces work.'

Y offers the book to X, who attempts to read, but can't, so improvises.

X Two be?
Or . . . *not* two be. Two be squared!
No . . . minus two B squared Y, plus X squared Y, plus Y cubed.

Y You're making theatre.

X Minus two B X Z, minus two X squared Z, minus two Y squared Z, plus Y Z squared equals zero.

Y That's great.

X A finite closed non-orientable manifold.

Y Keep going . . .

X tears a page out of the book and makes a Möbius strip from it. Y is horrified.

Y What are you doing?

X I'm physicalising space.

Y You mustn't do that! That's my Shakespeare.

X proudly shows Y the shape.

X A twisted space.

Y It's rubbish . . . pathetic. That's not beautiful.

X It's a prop? It's a strange loop.

Y It's a crappy little Möbius strip. And that helps us how?

X I wanted to make theatre. I thought it might help to find you OUT.

Y But instead you just ruined my most precious book.

X I'm sorry.

X tries to return the torn page to the book but it just falls to the floor.

Y That won't work.

X I didn't know it was a non-invertible function.

Pause.

Y It's alright. It is just a book.

Y places the Shakespeare standing up on the floor at the back of the cube and turns to X.

And I appreciate the effort.

X Oh . . . that's alright. I'm not . . .
I can't . . .
Y equals Victoria?

Y Yes.

X Can I ask you a question?

Y Yes.

X What is your OUT?

Y I don't know.
Something.
Not nothing.
That's what this story is all about. The search for happy ever after.
But it always ends in nothing.

X Nothing doesn't exist. It's just an idea.

Y It does.

X Doesn't.

Y Does.

X Doesn't.

Y Does.

X Doesn't.

Y Does.

X Doesn't.

Y (*with ominous gravity*) The heat death of the universe.
OBLIVION!
The end of time. That is NOTHING!

X Oh . . . That is nothing.

Y That's the ending waiting for us all . . .
At the end of ALL our scripts.

X Even the variables?

Y Every single one of us.

X So they're not out either.

Y No, they're not OUT.
They might have come out,
On a night out,
To come in here to try and find out their OUT.
But they're not OUT.
They're all just in another IN.

X Looking in, waiting for the end of their scripts, where there is nothing . . . nothing . . . nothing . . .

Y Looking in on shadows . . .

Y has picked up the model cube and holds it so that its centre coincides with that of X's cube. This creates the suggestion of a tesseract. Inside the model, Y performs a piece of theatre with the figure of Y on a stick.

'Out, out, brief candle!
Life is but a walking shadow, a poor player
That struts and frets his hour upon the stage
And then is heard no more: it is a tale
Told by an idiot, full of sound and fury,
Signifying nothing.'

X Signifying something.

Y No, I think you'll find its nothing.

X It signifies something more. A cube inside a cube.

Pause.

It's a shadow.

Y A shadow?

X A shadow of a four-dimensional shape.
If I were to shine a light on a cube in four dimensions, then the shadow I would get would be a cube inside a cube.

The shadow proves the existence of what is casting it, even if what is casting it is unseeable.

There must be something more.

Y If you say it's unseeable, that means you can't show it to me.

X Just because you can't see it, doesn't mean it doesn't exist.

We again hear the recurrent offstage train rumble.

Like that, that . . . sound.

We're never going to know what that is. That sound is like the shadow.

It shows the existence of something more, even though we'll never know what it is.

Y But I do know what that sound is.

X Really?

Y It's the underground.

X Underground.

Y Yes. It's the Circle Line.

X The circle . . . line?

Y Where the underground trains run.

X Trains? What's a trains??

Y A train . . . is a function that translates variables from one coordinate to another . . .

X (*laughs*) That's a lovely idea, dear, but you'll never be able to show it to me.

Y Yes, I can.

X Really?

Y If we went through that door at the back . . .

X Door at the back?

Y points to the exit at the back of the auditorium.

Y That door at the back . . . there . . . see . . . under the little green running man . . . see it?

X That's just another of your imaginary doors.

Y No, *that's* a real door.

X What's on the other side?

Y Well . . . there's the lobby of the Pit theatre, and then . . . if you go up the stairs and through the building, then after about an hour and half you'll be on the street, on Silk Street, where you can walk through that massive tunnel to the Barbican Underground station to get to the train.

X The street. That's OUT!

Y That's not OUT. That's just in another in. That's just a street, in a city, a place full of traffic lights . . . and cashpoints . . . and sick . . . and honking horns . . .

X Honking horns? That's OUT!

Y No, that's not OUT. That's just in another in. In the city.

X Can you get out of the city?

Y Yes, but then you'd be in the country. With fields of wheat . . . and wind farms . . . and in-breeding . . . and cows.

X Cows? That's OUT!

Y No, that's not OUT. That's just in another in. In the country.

X Can you get out of the country?

Y Yes, but then you'd just be in space, with lots of . . .
space . . .
and planets . . . and space . . .
and stars . . . and space . . .
and universes . . .

and dark matter . . . and space . . .
and wormholes . . .

X Worms in holes. That's OUT!

Y No that's not OUT. That's just in outer space. It doesn't matter how far you go, you can never get OUT, you're still just in another IN.

Pause.

X I want to go out there.

Y You can't.

X Why?

Y Because you're X. We made you up. You're a character in the script.

X I want to go out.

Y I'm afraid you're scripted to stay in here.
A manifold needs no ambient space in which to exist.

Pause.

X And you? What happens to you?

Y Oh, I'd love to stay in here.
I'd love to be in your OUT.
I'd love to lie here in the shadow of the four-dimensional cube knowing it exists without needing to see it. To be still.
But the script says that I keep going through room after room after room.

X Tomorrow and tomorrow . . .

X *and* **Y** and tomorrow . . .

Y Creeps in this petty pace from day to day,
To the last syllable of recorded time.

X What is the last syllable of recorded time?

Y cannot bear to say it.

What does the script say?

X retrieves the script from the white box.

Y Nothing.

X It can't say nothing.

Y It says NOTHING.

X finds the end of the play and reads.

X 'NOTHING.'

Pause.

I'm going to rewrite this script.

Y What?

X I'm going to give you OUT.

Y You can't.

X I'm going to give you *my* OUT.
Where there is something, not nothing.
Where numbers go on for ever . . .
Where . . .

X *and* **Y** Parallel lines meet . . .

Y Where things can exist without being seen.

X Where there are ideas that don't die,

Y Signifying . . . SOMETHING.
That's not possible.

They begin to circle in different directions round the cube. The energy builds.

The idea of the square root of minus one might go on forever, but how can 'I'?

X What is 'I'? Maybe 'I' is a piece of mathematics that can exist for ever.

Y I can't exist for ever. You can't just give me your OUT. And it's not possible to rewrite the script in the middle of a show.

X stops pacing.

(*Sotto.*) Although I've completely lost track of what's possible and what's real . . .

X Y!

Y stops pacing. They are both still.

(*Calmly.*) Change the axioms. Break the rules.

Y We can't.

X This is theatre. Anything is possible. You taught me that. Stand there. Don't move. Hold this. In both hands. Very carefully.

X hands Y the script which we see is made from many A4 pages Sellotaped together to make one long strip.

I'm going to take the end of the script . . . We're going back to the beginning . . .

X takes one end of the long script which unfolds itself as Y holds the other end.

Y I always go back to the beginning, and the ending never changes.

X Not this time.

X goes to attach their end of the script to Y's, beginning to make a loop, but at the last moment twists it before joining both ends together.

Y You're putting a twist in the script.

X It's a Möbius script.

Y laughs.

Oh, I made a joke.

Y Yes, you did.

X indicates the 'Möbius script'.

X Theoretically, if something goes round this universe, then it gets flipped over.

Y A strange loop. So theatrically . . .

X A change of variable.

Y Then Y becomes X.

X And I become U.

Y carefully drapes the Möbius script around X's neck, dressing X ritualistically. Y places the white cube box and the bag on X's shoulders before finally handing X the cube seat to carry. X now looks like Y did at the beginning of the play with the addition of the Möbius script.

Y Are you sure you want to do this? As I tends to infinity, U tends towards zero. No more plus one?

X The street.
Honking horns.
Cows.
Worms in holes!
OUT.

A laden X moves to the downstage edge of the cube and hesitates. Unseen by X, Y once again seems to infuse X with the courage to go as Y mimes a final push from behind. X steps out of the cube and begins to move excitedly towards the exit.

Y Y!!!

X stops, suddenly realising it is they who are being called. X goes back to the cube, where Y hands over the precious Shakespeare.

Everything you need is in here. Look after it.

X takes the book.

X Everything *you* need is in my cube.

X, who is now Y, turns and leaves through the audience exit.
Y, who is now X, remains in the cube alone.
Y closes their eyes and begins a Platonic Sequence of movements – similar but not identical to those performed at the beginning of the play. Y finishes and opens their eyes.

Y OUT.

A Mathematical Prompt Book

In theatre, the prompt book contains the set of instructions used by the company to physically construct the play from its elements. In mathematics, formulae and proofs are sets of instructions used to coordinate the relationships between objects within the mathematical space. *I is a Strange Loop* has a mathematical prompt book: a set of instructions that bring its formulae and proofs to life as elements within the play.

DRAMATIS DEMONSTRANDAE

First Platonic Sequence
The ruler and compass construction of the hexagon

The Ancient Greeks enjoyed the challenge of constructing geometric shapes using straight lines and arcs of circles drawn using a straight edge (with no measurement) and a pair of compasses. One of the shapes that it is possible to construct is a perfect hexagon. X's first Platonic Sequence physicalises this construction in a vertical plane facing the audience. It's as if there is a blackboard running through the middle of the cube.

The sequence starts with a horizontal straight line that X traces out from stage-right to stage-left. X then bisects the line. This is achieved by arcs of two circles each centred at the two ends of the line traced out by X, one hand fixed at the end of the line, the other hand sweeping out the arc of a circle. The radius of each circle should be the same, roughly the span of X's outstretched arms. This will be more than half the length across the cube ensuring that the two arcs intersect in positions roughly at head and groin height. X then places hands at the two points in space where the arcs

63

of the circles intersect and draws a line by joining the hands together. The hands meet at the midpoint of the first line drawn.

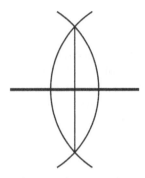

X sweeps out a complete circle using this point as the centre. The first line X drew sits now as a diameter across the circle. Taking one of the points at which the diameter meets the circle as the centre of a new circle, X sets the other end of the pair of imaginary compasses to the centre of the first circle and sweeps out an arc of this new circle so that it intersects the first circle in two new points. The move is mirrored with the other end of the diameter as the centre of a third circle.

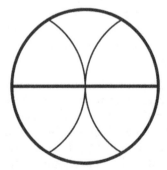

There are now six points marked on the first circle: the two ends of the diameter and the four points marked by the arcs of the two circles drawn. These points divide the circle

perfectly into six equal parts. X now connects the points in
pairs to create a perfect hexagon.

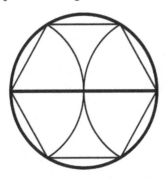

Second Platonic Sequence
Geometric proof of the irrationality of the square root of 2
using infinitely decreasing squares

The Ancient Greeks proved it is impossible to find a square
whose sides and diagonal both have lengths which are
whole numbers. Suppose it was possible and the sides of the
square have length q with diagonal of length p. If you draw
a new square in the centre of this square whose corners are
at a distance of (p–q) from the corners of the original
square then you will find that the diagonal and the sides of
this new square also have lengths which are whole
numbers. These whole numbers are smaller. But you could
keep doing this making smaller squares whose sides and
diagonals are whole numbers. The squares can go on to
infinity but decreasing whole numbers don't. This
contradiction implies the original square with whole
number side and diagonal is impossible to construct.

Because the length of the diagonal across the square is √2
times the length of the side of the square, this implies that
there is no fraction whose square equals 2. The square root
of 2 was a new sort of number, called 'irrational' because it
wasn't the ratio of two numbers.

In the second Platonic Sequence, X physicalises this proof. X starts by drawing a large square in the vertical plane facing the audience. X then draws a diagonal line from the top left corner to the bottom right corner. X then marks out the arc of a circle of radius the length of the side of the square. The circle is centred on the top left corner and starts at the bottom left corner. The arc will intersect the diagonal X drew across the square. X now draws a new smaller square inside the larger square. The first corner of the new square is the point at which the arc intersects the diagonal.

X then repeats the same sequence for this new smaller square. X goes on creating a sequence of concentric squares getting smaller and smaller. If the original square had sides and diagonal with whole number lengths this sequence of squares would terminate. X realises that the sequence of squares never terminates, implying that the square root of 2 cannot be written as a fraction.

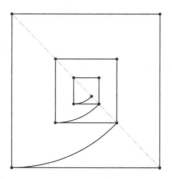

X and Y's First Duel

Suppose

$$Y = 1$$

Multiply both sides by Y

$$Y^2 = Y$$

Subtract 1 from each side

$$Y^2 - 1 = Y - 1$$

Factorise the left hand side

$$(Y + 1)(Y - 1) = Y - 1$$

Cancel the common factor $(Y - 1)$

$$(Y + 1) = 1$$

Therefore

$$Y = 0$$

The false move is cancelling the common factor from line 4 to 5. Since $Y = 1$, this means you are dividing by $Y - 1 = 0$. You're not allowed to divide numbers by 0.

X and Y's Second Duel

Suppose

$$X = 1$$

Multiply both sides by X

$$X^2 = X$$

Differentiate with respect to X

$$2X = 1$$

Therefore

$$X = ½$$

How did Y manage to cut X in half? The false move is differentiating with respect to X. You can only differentiate with respect to a variable but $X = 1$ is constant and not a variable.

Proof by Induction

Proof by Induction allows the finite mathematician to navigate the infinity of mathematics. You might have a formula that you want to prove works, whatever number you insert into the formula. But how can you check it for the infinite choice of numbers that mathematics offers? The proof by induction works on the principle of how to get someone to climb an infinite ladder. First you show them how to get onto the first rung of the ladder. Then you show them how if they've got to the nth rung they can get to the next rung up. It's the combination of these two instructions that allows someone to climb as high as they want.

Proof by induction that

$$1 + 2 + 3 + \ldots + n = n(n + 1)/2$$

First check that the formula works for $n = 1$. The formula $n(n + 1)/2 = 1(1 + 1)/2 = 1$. So the formula is correct for $n = 1$.

Now suppose the formula is correct for all n from 1 to N. We now check that the formula works for $n = N + 1$.

But we know the formula works for adding up the numbers to N so let's substitute this formula into the equation:

$$(1 + 2 + 3 + \ldots + N) + (N + 1)$$
$$= N(N + 1)/2 + (N + 1)$$
$$= (N^2 + N + 2N + 2)/2$$
$$= (N + 1)(N + 2)/2$$

which is indeed the formula with $n = N + 1$. Having proved the formula works for $n = 0$ and also proved that if it works for $n = N$ it works for $n = N + 1$, we now have all the inductive steps in place to climb our infinite ladder and prove that the formula works for all n.

X

A variable.

Y

A(nother) variable.

The Parabola

Galileo discovered that the path traced out by an orange thrown through the air is given by a quadratic equation called a parabola. Suppose the orange is launched such that the speed in the horizontal direction is u and speed in the vertical direction is v. If Y is the horizontal distance that the orange has travelled then the height Z in the vertical direction from the point it was launched is given by the following equation

$$Z = (v/u)Y - (G/2u^2)Y^2$$

where G is a constant equal to the acceleration towards the floor. X chooses to capture the second orange using the values $v = 1$ and $u = 1$.

Three-Dimensional Spiral

Each point in three-dimensional space can be described by its coordinates (X,Y,Z). The coordinates tell you how far to move in each direction to get to the point. For example, the point $(1,2,3)$ is got to by moving 1 step in the X direction, 2 steps in the Y direction and 3 steps in the Z direction.

A clever way to describe a path traced out in three dimensions is to use a fourth variable t called a parameter. You can think of t like time. At time t the equations tell you where you are in three-dimensional space. For example, X uses the equations for a three-dimensional spiral to describe a path for the third orange to follow, eventually arriving in X's hand located at point $(0,0,0)$. After t seconds the coordinates (X,Y,Z) of the orange are as follows:

$$X = e^{-t} \cos t$$

$$Y = e^{-t} \sin t$$

$$Z = 1/t$$

The first two equations tell us that the orange is following a logarithmic spiral path and the third equation tells us that the height above X's outstretched hand is getting smaller and smaller. In theory the orange only arrives at $(0,0,0)$ when t reaches infinity.

The Sphere and the Ellipsoid

The sphere is the shape where each point is an equal distance R from the centre. Suppose the centre is located at the point $(0,0,0)$. Then the distance from the point (X,Y,Z) to the centre is given by the square root of $X^2 + Y^2 + Z^2$. So if we take the shape whose coordinates (X,Y,Z) all have the property that $X^2 + Y^2 + Z^2 = R^2$ where R is some fixed number, then all these points are a fixed distance R from the centre at $(0,0,0)$. So the equation

$$X^2 + Y^2 + Z^2 = R^2$$

describes the points (X,Y,Z) lying on a sphere of radius R.

An ellipsoid is a sphere which has been pushed and pulled in three different directions. It is the shape of a rugby ball. We can squeeze and stretch the sphere by scaling the coordinates X, Y and Z. The equation

$$(X/A)^2 + (Y/B)^2 + (Z/C)^2 = R^2$$

describes the points (X,Y,Z) lying on an ellipsoid. If you cut an ellipsoid the shape you get is an ellipse.

A Genus One Surface

The mathematics of topology regards two shapes as the same if one shape can be moulded into the other without cutting it. Topology is often called bendy-sheet geometry. For example you can mould a sphere into an ellipsoid without cutting it. A bagel can be moulded into the shape of a coffee cup with a handle. But you cannot mould the bagel into a sphere without cutting it. The French mathematician Henri Poincaré proved that every 3D shape whose surface is finite in area can be moulded into a sphere or a bagel with one hole, or a bagel with two holes or a bagel with more holes. The number of holes is called the genus of the shape. A bagel is therefore a genus one surface. A teapot is a genus two surface. A pretzel is a genus three surface.

The mathematical name for a bagel is a torus. The torus is a circle's worth of circles. Think of a circle of radius R_2 which is then swept out in a second circle of radius R_1. The inner tube of a bicycle is a torus. The following equation describes the coordinates of a torus in three-dimensional space:

$$(R_1 - \sqrt{X^2 + Y^2})^2 + Z^2 = R_2^2$$

where R_1 is the radius of the centre of the torus to the centre of the tube and R_2 is the radius of the circular cross-section of the tube.

How can you tell if you're on the surface of a sphere or a torus without flying off the surface and looking from the outside to see if there is a hole? Start two journeys round a sphere from a fixed point and they will cross each other on the other side of the sphere. But on the torus, you can make two journeys round the surface that never cross each other.

In the computer game Asteroids the universe consists of the finite computer screen. But this finite universe has no walls. If you travel off the top of the screen you reappear at the bottom. Head off to the left and you reappear at the right. To understand the shape of this universe, first connect the top and bottom of the screen to make a cylinder. Now join the left and right to create a bagel shape or a torus. This is the shape of the universe in the game of Asteroids. It is also the shape X suggests for the universe of the play. Exit stage-left you re-enter stage-right. Exit through the ceiling of the cube you re-enter through the trap door in the floor.

When X exits through the audience at the end of the play there is the possibility that this direction is also looped. Perhaps X will eventually find themselves returning to the stage through the flap at the back. These three non-intersecting loops would imply that the universe of the play is the three-dimensional surface of a four-dimensional bagel.

The Möbius Strip

To construct your own Möbius strip, cut out the strip on the side. Now join the ends up in a loop but before you join the ends, make a twist of 180 degrees in one end (join X to Y and Y to X). The Möbius strip, discovered in 1858 by German mathematician August Möbius, has some very curious properties. The shape has only one edge and only one side. Take a pair of scissors and cut the shape down the line running round the strip. Instead of the two loops of paper that you'd expect by cutting something in half, the result is a single piece of looped paper with two twists in it.

Make the Möbius strip from clear plastic and place a cut-out of a right hand on the plastic. Now move the hand round the strip. When it returns to the position it started the hand has flipped over to become a left hand. This is why we call the surface non-orientable. It flips things over.

The following equation can be used to create a Möbius strip:

$$-2B^2Y + X^2Y + Y^3 - 2BXZ - 2X^2Z - 2Y^2Z + YZ^2 = 0$$

A Geometric Series

The Ancient Greek philosopher Zeno was rather perplexed as to how an arrow ever reaches its target. Suppose the target is 16 metres away. First the arrow must travel half the distance to the target: 8 metres. Then half the distance left: 4 metres. Then half the distance left again: 2 metres. The arrow seemingly has to cover infinitely many distances, each half the length of the previous distance. How can it ever achieve infinitely many actions in a finite time? Suppose the arrow travels at a metre per second (a very slow arrow but it helps the maths). How long does it take to cover these infinite number of steps to hit the target?

We need to add up

$$8 + 4 + 2 + 1 + \tfrac{1}{2} + \tfrac{1}{4} + \ldots$$

Call this infinite sum X. If we double each term in this infinite sum then

$$2X = 16 + 8 + 4 + 2 + 1 + \tfrac{1}{2} + \tfrac{1}{4} + \ldots$$

Now take X away from this sum

$$2X - X$$

$$= 16 + 8 + 4 + 2 + 1 + \tfrac{1}{2} + \tfrac{1}{4} + \ldots$$

$$- (8 + 4 + 2 + 1 + \tfrac{1}{2} + \tfrac{1}{4} + \ldots)$$

$$= 16$$

because every term except 16 gets subtracted by a term in the second infinite sum. But $2X - X = X$. So

$$X = 16.$$

It takes 16 seconds for the arrow to cover these infinitely many steps and hit the target.

Nothing

Zero is a relatively new number. It was discovered in the seventh century by the Indian mathematician Brahmagupta. Before that people didn't think you needed a number to count nothing. Indian mathematicians did calculations with stones in the sand. The symbol 0 might have its origins in the shape that was left in the sand when there were no stones. Brahmagupta deduced rules for calculating with zero but he came a cropper when he tried to understand dividing by 0.

Infinity

For centuries infinity was regarded as a concept beyond the conception of the finite human mind. But at the end of the nineteenth century Georg Cantor found ways to navigate the infinite. He proved that there are many sorts of infinity . . . infinitely many, some bigger than others. Leopold Kronecker thought Cantor's mathematics an aberration but David Hilbert came to his defence: 'No one shall expel us from the paradise that Cantor has created for us.'

Singularity

Singularities are points at which mathematical functions crash. A function in mathematics is like a computer program. You input numbers, the function calculates away and then spits out an answer. The function has a singularity if the calculations crashes or doesn't make sense. For example the function $1/X$ outputs the answer infinity if you put $X = 0$. A black hole is a point in space where gravity becomes infinite. It is a space-time singularity.

Manifold

A manifold is a shape that can be broken up into small pieces that individually look like a flat piece of paper (or the analogue in higher dimensions). For example the 2D surface of a sphere is not flat but you can divide it up into bits that individually look flat. Think of the way a football is made from flat pieces of leather shaped like pentagons and hexagons. Our three-dimensional universe is an example of a 3D manifold where each small piece of the universe can be described by a 3D grid (the 3D analogue of a flat piece of paper). If our universe is finite, how are these pieces of 'flat' space put together? What shape does the universe have? Is it the surface of a 4D sphere or perhaps a 4D torus? The interesting point is that a 3D manifold can exist without having to live in a larger 4D space. It just makes it easier to imagine. A manifold needs no ambient space to exist.

Tesseract
A cube in the fourth dimension

If you want to depict a three-dimensional cube on a two-dimensional piece of paper then you might draw a small square surrounded by a larger square and then connect the corners. Suddenly a shape with depth appears on the page. To create a shadow of a four-dimensional cube in three dimensions you can play the same game. Build a small three-dimensional cube inside a larger cube and join the edges. You have built a 3D shadow of the four-dimensional tesseract.

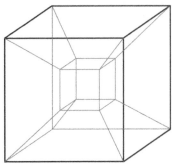

A shadow of a 4D cube

Strange Loop

What makes a loop strange? A strange loop is a cyclical structure that goes through several layers of a hierarchical system only to find itself back at the beginning. One layer is contained inside another only for there to be a strange crossing of these levels so that the higher level suddenly appears embedded in the lower level.

Some of Escher's most famous images capture visually what a strange loop looks like. Think of the hand drawing the hand drawing the first hand. The image of the hand on one side of the picture is being drawn by a hand that appears to be higher up in the hierarchy. But then this hand we find is being drawn by a hand that surely exists in an even higher part of the hierarchy. But then we realise that this is the hand that we started with. Another example is Escher's image of the monks climbing a staircase that loops round on itself so that they are always climbing. It turns out that Bach's compositions too exploit these strange loops to create interesting musical structures. The mathematician Kurt Gödel showed how a sufficiently complex system might use a logical strange loop to talk about itself.

This is what inspired Douglas Hofstadter in his book *Gödel, Escher, Bach* to propose that the concept of a strange loop might be the key to creating consciousness, the brain's ability to conceive of itself.